Introduction

Easy to Make! Easy to Read!

Young children get excited reading about things that relate to their everyday experiences. But they get even more excited when they can help create the art and text in these reading materials. The books in the *Make Your Own Emergent Readers* series provide children with interactive, personalized reading experiences by inviting them to investigate, complete sentence frames, illustrate, put together, and create covers for their very own readers. Each book in the series contains 10 eight-page themed reproducible books for emergent readers. These easy-to-read Little Books contain text that is simple, predictable, and repetitive, and usually have one or two new words or changes on each page. The books present various subject matter following a specific theme.

This resource, *Little Science Stories,* provides everything you need to get children excited about reading as they create their very own library of self-made and illustrated science books.

This book includes:

Creative Suggestions and Activities

Pages 3–7 provide creative suggestions and activities for each Little Book, including:

Cover Idea—Tells you how to create a colorful, eye-catching cover using a variety of art techniques and mediums.

Ideas for Illustrating Inside Pages—Helps you guide children to add their own text to many of the books and creatively illustrate the pages.

Extension Activities—Provide unique ways to extend learning concepts presented in the Little Books as well as encourage further reading and writing.

Literature List

A comprehensive list of related books is provided on page 8. Use these books as read-alouds, research material, and extra books children can read on their own. Theme-related read-alouds enhance your literacy program by providing background knowledge and extending the concepts introduced in the content areas.

Directions for Making the Little Books

Putting the Pages Together

Reproduce the pages for each Little Book. *Do not cut pages in half!* Children can fold the pages and place them in numerical order. The pages will be doubled, and text will read on both sides. This unique page construction creates more durable books for frequent rereading. Also, creating artwork is easier. Markers and paint will not bleed through, and the pages are more suitable for collage materials.

Note: Depending on the art technique used, it may be easier for children to illustrate each page before assembling books, especially if it's a messy technique! Have extra pages available for "mess-ups." Once the art is complete and pages are in order, fold the pages on the dotted line and bind them together. Stapling is the easiest, but pages can also be hole-punched and bound with ribbon or yarn.

Adding a Cover and Completing the Pages

- The first page of each Little Book can serve as the cover. However, you may choose to follow the cover suggestions on pages 3–7, or invite children to each make a special, individualized cover of his or her own. Make two covers at a time by cutting 12" x 18" (30 cm x 45 cm) construction paper in half lengthwise, and then folding each piece in half.
- The inside pages can be illustrated quickly and easily by drawing and/or coloring. (Refer to the Literature List on page 8 for books that help children learn to draw.) Or, you may follow the suggestions on pages 3–7 for more visually-exciting art. In many instances, children are asked to complete existing art and/or draw inside borders or frames. These suggestions are a great way to integrate the arts into your program, and children will love experimenting with collage materials, pastels, paint, crayon-resist, cut paper, and more!
- In advance, make several photocopies of your class picture. Children can use their own pictures whenever a self-portrait or photo is called for in the art. Or, you can use photocopied pictures to create a "Meet the Author" page for special books.
- You may also attach a "comments" page on the inside back covers of the books. Here children can dictate or write their feelings about the book, what they learned, or more text additions to continue the story.
- To encourage rereading, attach a page to the inside back covers on which children can list the people to whom they have read their books.

Using the Little Books

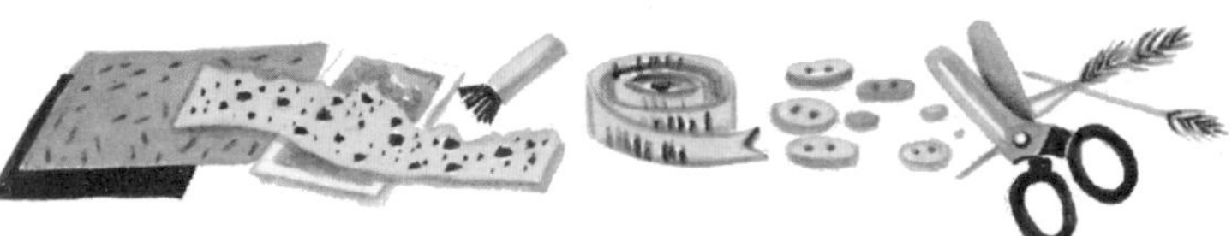

Before children make and read Little Books on their own, introduce them to the text during shared reading activities. Depending on children's reading-skill development, they should have several opportunities to interact with the text before making each Little Book. You can do this in a variety of ways:

- Make a class big book using the text from the Little Book combined with student illustrations. Read the big book several times during shared reading time.
- Make a pocket chart of the Little Book's text, along with matching picture cards.
- Make overhead transparencies of the Little Book's text and art.

Just think of how proud children will be showing off their own creatively-designed books to their classmates and families!

There are a myriad of ways to use these books in your classroom and at home, but here are just a few suggestions on how to incorporate the Little Books into your literacy program:

- Learning-center activities
- Extra reading in the content areas
- As an introduction to various subjects
- Independent reading material
- Homework activities
- Reading material for the child at home

Creative Suggestions and Activities

Hello Day! Hello Night!

Cover Idea

Use a blue construction-paper cover. Divide the front in half by attaching a black square of construction paper. On the blue side, add a big, bright yellow sun to represent day. On the black side, add a white moon and star shapes to represent night. Or, paint these shapes using yellow and white tempura. Write *Hello Day!* on the blue side and *Hello Night!* on the black side.

Inside Pages

In the empty frames on each page, draw and color pictures of things you do during the times described. On page 7, draw yourself in bed and attach a piece of fabric to the bedspread. Draw a pet cat or dog sleeping on the rug, and objects on the bedside table to reflect your own room. Color page 8 and look for the nocturnal animals. Then add more nocturnal animals to the picture with drawings or cutouts.

Extension Activities

- Discuss what causes day and night, and how the earth revolves around the sun. Then investigate different phases of the moon. Invite children to draw on a calendar what the moon looks like each night.
- Create a Venn diagram illustrating daytime activities, nighttime activities, and activities done both during the day and at night.

What Does a Seed Need?

Cover Idea

Use a blue cover or sponge-paint a white cover with blue tempera. Add a brown flowerpot cut from a recycled lunch bag. Glue dried coffee grounds to the flowerpot as "soil," and add a dried lima bean. Attach fabric or wallpaper scraps to represent curtains in a window. Add a yellow sun peeking from behind the curtains. Write the title on the flowerpot.

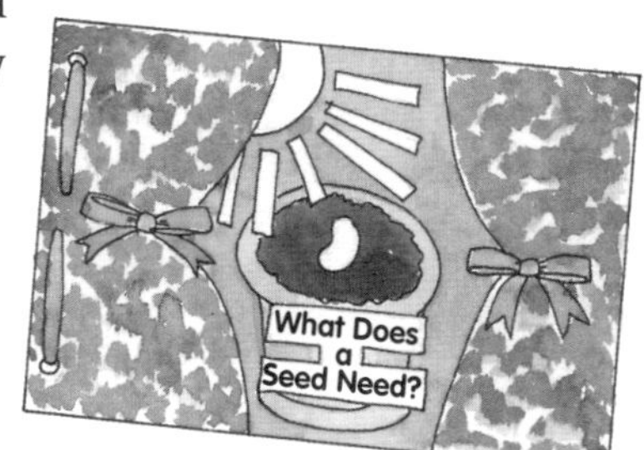

Inside Pages

On page 2, "paint" glue to the pile of soil. Sprinkle dried coffee grounds over the glue. On page 3, decorate the watering can. Cut water drops from construction paper or aluminum foil and glue them "falling" from the can. On page 4, watercolor a blue sky and add wispy pieces of cotton as clouds. On page 5, attach a sun cut from yellow construction paper or gold foil. On pages 6 and 7, draw a slowly-sprouting flower plant in the pot. On page 8, draw a fully-bloomed flower. Add green tissue paper to the leaves and bright colors to the flower petals.

Extension Activities

- Create a classroom garden. Grow a variety of plants from seeds, or place clippings from houseplants in water until they root. Plant seeds from raw fruits and vegetables eaten during a tasting party to predict and then observe which ones sprout. Plant "mystery seeds" from a variety of packages. Place the packages on display so young botanists can observe and guess which seeds are growing.
- Design a scientific experiment proving the facts in the Little Book. For example, for *A seed needs water,* plant a seed in soil in a plastic cup. Give the seed sunlight and air, but no water. Have children observe and record results from each experiment.

Living Things Grow

Cover Idea

Decorate the cover with a collage of pictures of living things (plants, animals, and people). Cut pictures from magazines, draw and color your own pictures, or use a combination.

Inside Pages

On page 1, draw a butterfly on the leaf, showing how a caterpillar grows. Add flowers to page 2 and trees to page 3. On page 4, draw baby birds in the nest. On page 5, draw a frog on the lilypad, demonstrating the final stage in a frog's life cycle.

Pages 6 and 7 are open-ended. Finish the text and pictures to record the changes in living things happening in the school environment (e.g., silkworms) or to reflect recent science lessons. On page 8, illustrate yourself or attach a baby picture along with a current photo.

Extension Activities

- Learn the song "Everything Grows" by Raffi, and share the accompanying picture book (see Literature List, page 8). Make a class book using the lyrics from the song or create a class innovation with your own new lyrics.
- Read *Is Your Mama a Llama?* by Deborah Guarino. Invite children to compare the baby animals to their grown mothers. Set up a "research center" with animal books and find which baby animals have special names. For example, a baby owl is called an "owlet." Make a big book based on children's findings, following a pattern such as:

  A calf grows up to be a (turn page) *COW!*

  An owlet grows up to be an (turn page) *OWL!*

Animal Builders

Cover Idea

Use a blue construction-paper cover. Create a tree branch with torn pieces of brown construction paper, and a nest resting on the branch with lighter brown paper. Add torn green tissue-paper leaves. Cut out or draw a bird in flight and add it to the scene. Using Tacky glue, attach small twigs, leaves, pieces of string or yarn, and other collage materials to add texture to the nest.

Inside Pages

On page 2, draw a spider on the web. On page 5, draw bees around the hive. Discuss and display books, nature magazines, and other materials about animals and the homes they build. Use the information to finish the text and art on pages 6 and 7. On page 8, draw a picture of something you built or attach a photograph of yourself with that object.

Extension Activities

- Display a variety of pictures of different bird nests or, if possible, an actual bird nest. Examine the materials birds use to build their nests. Gather twigs, grass, leaves, mud (or clay), string, and brown yarn pieces, and invite children to experiment with building their own bird nests! Try building them with tweezers (as "bird beaks"). Compare finished nests with those the birds built.
- Discuss how each animal's home suits its needs. Imagine if a bird tried to live in a spider's web, or if a bee tried to live in a bird's nest! Have children tell about how their homes are "just right" for their needs.

A Lot of Energy!

Cover Idea

Sponge-paint a small paper plate yellow and add sunrays cut from yellow construction paper. Attach the finished sun to your cover.

Inside Pages

On each page, draw yourself performing the described activity. On pages 4 and 7, draw yourself doing a favorite high-energy activity. You can also attach photos. On page 8, draw pictures of healthy foods or attach cutouts of fruits and vegetables from supermarket advertisements.

Extension Activities

- Make a class graph with three columns titled *Low Energy, Medium Energy,* and *High Energy.* On small paper squares, illustrate activities the class experiences throughout the day. Add your pictures to the appropriate columns of the graph.
- Use this opportunity to teach children about healthy food choices and a properly balanced meal. Have them cut pictures from magazines of meat, vegetables, fruit, bread, and so on, and glue them to a paper plate. Display these healthy "meals" on a bulletin board titled *The Energizers!*

I Learn with My Five Senses

Cover Idea

Draw part of the earth on the bottom of the cover and paint it blue and green. Decorate a self-portrait paper doll and glue it on the earth. Add pictures/stickers of everyday things you see, smell, feel, taste, and hear.

Inside Pages

Complete the text and art after a planned "five senses" experience such as a popcorn party, trip to the park, or seasonal walk. For example:

I learn by seeing. I can see <u>red and gold leaves</u>.
I learn by hearing. I can hear <u>wind in the trees</u>.

On page 8, complete the park scene showing things that can be experienced by seeing, hearing, smelling, touching, and tasting.

Extension Activities

- Add an extra blank page between each page of the Little Book. On these pages, have children describe the sight, sound, smell, touch, and taste experienced. For example:

  I learn by seeing. I can see a rainbow. It is colorful.

  I learn by smelling. I can smell a cupcake. It smells like chocolate.
- Have children experience just how important their senses are. Peel an apple and a potato, and cut them into tiny pieces. Have children hold their noses and try a piece of each. Can they tell which is which? Have them release their noses and try again. Discuss how we rely on our senses to keep us safe—what if some meat looked fine, but smelled rotten . . . would we eat it? How do our eyes and ears keep us safe when we're walking down a busy street? Write a class book on different "sense" experiences. For example, *I can see when the light turns green and it's safe to cross the street.*

What Am I?

Cover Idea

Glue a large black question mark to a yellow construction-paper cover. Around the question mark, place animal stickers, magazine cutouts, or decorate with rubber animal stamps.

Inside Pages

Finish each picture by drawing the appropriate animal in each habitat. Have a variety of books about animals available for reference. Cover the animal on each page with a construction-paper flap labeled with a question mark. When sharing your book, have your partner guess the animal, then lift the flap. Finish page 8 with a description of yourself. Make the flap in the shape of a house. Place a photo of yourself under the flap.

Extension Activities

- Choose one of Earth's habitats for study. Create a class guessing book featuring plants and animals from that habitat.
- Create "I Spy" games by gluing large magazine pictures of animals onto 9" x 12"(22.5 cm x 30 cm) construction paper. For each animal, obtain another sheet of construction paper. In each, cut a 2" x 2" (5 cm x 5 cm) flap. Position the flaps in different places. Paper-clip the paper with the flap on top of an animal picture. Invite children to try and guess what the animal is using the visual "clue" under the flap.

Hooray for the Earth!

Cover Idea

Cut a blue construction-paper circle to fit a black cover. Finger-paint continents on the circle with green paint, leaving a lot of blue showing to represent oceans. Add a few wispy pieces of cotton for clouds. Glue the painted earth shape to the cover. Add gold sticker stars around the earth.

Inside Pages

Add to each picture by drawing mountains, flowers, trees, and so on. Outline the prominent designs with white glue. Allow to dry overnight. Color within the glue lines with Cray-pas (oil crayons), colored chalk, or pastels. Use Tacky glue to add a ribbon to the earth on page 6. On pages 7 and 8, draw two ways we can care for our planet and home.

Extension Activities

- Learn the song "Hooray for the World" by Red Grammer (see Literature List, page 8).

  Hooray for the world, I'm glad to be on it!
  Hooray for the world, I'm glad to be on it!
  Hooray for the world, it's a special place.
  We've got mother nature and the human race.

- To extend the ecology theme, create a class book in the shape of the earth. Illustrate in text and art some good ways to care for our planet Earth. For example:

  We can pick up trash.
  We can recycle glass and plastic bottles.
  We can plant trees.

I Like Trees

Cover Idea

On a blue cover, form a tree trunk with torn pieces of recycled lunch bags. Add leaves, flowers, and fruit reflecting your favorite season with colored tissue paper (green leaves and pink flowers for spring; red, orange, and yellow leaves for fall).

Inside Pages

Use a combination of crayon and watercolor to finish the art. Outline the trees with black crayon before painting inside the lines. Draw your favorite tree in the park on page 3. Draw yourself on the swing on page 4. Sponge-paint seasonal trees on pages 6 and 7. Draw yourself sitting in the branches of a winter tree on page 8.

Extension Activities

- Read the book *A Tree Is Nice* (see Literature List, page 8), then put on a play! Have several children "decorate" themselves as trees and stand in the background as other children come up one at a time to tell why "a tree is nice." Use simple props such as sunglasses *(a tree gives shade on a sunny day)*, or a book *(under a tree is a great place to read a book)*. Put on a play for other classes or parents on Arbor Day or Earth Day.
- Go on a leaf-gathering expedition! Give each child a brown lunch sack in which to collect their leaves. Bring paper, writing boards, and crayons with you on the walk. When you arrive at a tree, have children sit around it and just observe. Invite children to draw the tree as realistically as possible. They can also make "bark rubbings" to experience the tree's texture. Back in the classroom, use the leaves children collected to make "leaf rubbings." Compare and contrast different leaves from different trees.

Where Does Water Go?

Cover Idea

Cut a large raindrop shape from light blue construction paper and glue it to a white cover. To make the raindrop look more "water-like," cut an identical raindrop shape from clear or blue plastic wrap and glue it over the raindrop.

Inside Pages

Draw pictures to match the text on each page with colored pencils or crayon. Using a marker, add arrows to each picture to show where the water is going. Children can add "mood" to their pictures with watercolor (gray for a stormy day; blue for a sunny day).

Extension Activities

- This book tells the story of the water cycle. Retell this never-ending circle story by illustrating it on a paper plate divided into eight pie-shaped sections. In each section, illustrate and write the text of each step in the water cycle. Ask children to give other examples of where they see the water cycle in action (e.g., a steaming pot on the stove; condensation on the outside of a glass of ice water).
- Invite children to discover the importance of water in their lives by making a daily log of every time they use water. Have them start from the moment they brush their teeth to when they call, "Mom, I need a glass of water!" from their beds at night. Or, consider making a class log of water use. Use chart paper and appoint a helper to write down the times and ways in which water is used.

Literature List

Air by David Bennett (Bantam)

And So They Build by Bert Kitchen (Candlewick)

Animal Architecture by Jennifer Dewey (Orchard)

Animal Fact, Animal Fable by Seymour Simon (Crown)

Animal Homes by Tammy Everts and Bobbie Kalman (Crabtree)

Animals and Where They Live by John Feltwell (Dorling Kindersley)

Apple Picking Time by Michele Benoit Slawson (Crown)

Beneath the Waves by Norbert Wu (Chronicle)

Cactus Hotel by Brenda Guiberson (Holt)

Eating the Alphabet: Fruits and Vegetables from A to Z by Lois Ehlert (Harcourt Brace)

Everything Grows by Raffi (Crown)

Experiments with Water by Ray Broekel (Childrens Press)

Fireflies in the Night by Judy Hawes (HarperCollins)

From Seed to Plant by Gail Gibbons (Holiday House)

Goodnight, Goodnight by Eve Rice (Greenwillow)

The Great Kapok Tree by Lynne Cherry (Harcourt Brace)

Growing Colors by Bruce McMillan (Mulberry)

Growing Vegetable Soup by Lois Ehlert (Harcourt Brace)

The House from Morning to Night by Daniele Bour (Kane-Miller)

I'm Growing by Aliki (Greenwillow)

Jack's Garden by Henry Cole (Greenwillow)

Just a Dream by Chris Van Allsburg (Houghton Mifflin)

The Midnight Farm by Reeve Lindbergh (Dial)

My Five Senses by Aliki (Greenwillow)

A Night in the Country by Cynthia Rylant (Macmillan)

Recycle! A Handbook for Kids by Gail Gibbons (Little, Brown)

A River Wild by Lynne Cherry (Harcourt Brace)

A Seed Is a Promise by Claire Merrill (Scholastic)

Silkworms by Silvia A. Johnson (Lerner)

Somewhere Today by Bert Kitchen (Candlewick)

Sun Up, Sun Down by Gail Gibbons (Harcourt Brace)

The Sun's Day by Mordicai Gerstein (HarperCollins)

A Tree Is Nice by Janice May Udry (HarperCollins)

Vegetable Garden by Douglas Florian (Harcourt Brace)

Water's Way by Lisa Westberg Peters (Arcade)

Welcome to the Green House by Jane Yolen (Putnam)

What the Moon Saw by Brian Wildsmith (Oxford)

Where Does the Sun Go at Night? by Mirra Ginsburg (Greenwillow)

Where the Forest Meets the Sea by Jeannie Baker (Greenwillow)

You Can't Smell a Flower with Your Ear: All About Your Five Senses by Joanna Cole (Scholastic)

You'll Soon Grow into Them, Titch by Pat Hutchins (Greenwillow)

Drawing Books

Draw 50 Series by Lee J. Ames (Doubleday)

Ed Emberley's Drawing Book: Make a World by Ed Emberley (Little, Brown)

Ed Emberley's Drawing Book of Animals by Ed Emberley (Little, Brown)

Ed Emberley's Drawing Book of Faces by Ed Emberley (Little, Brown)

Ed Emberley's Great Thumbprint Drawing Book by Ed Emberley (Little, Brown)

Ed Emberley's Picture Pie: A Book of Circle Art by Ed Emberley (Little, Brown)

I Can Draw Series by Lisa Bonforte (Simon & Schuster)

Hello morning!

Hello Day! Hello Night!

Hello Day! Hello Night! written by Rozanne Lanczak Williams

Hello day!

3

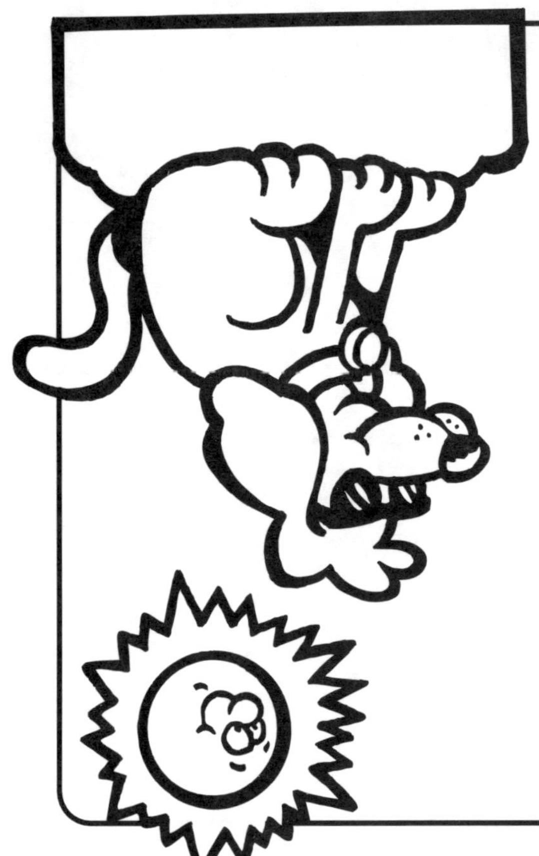

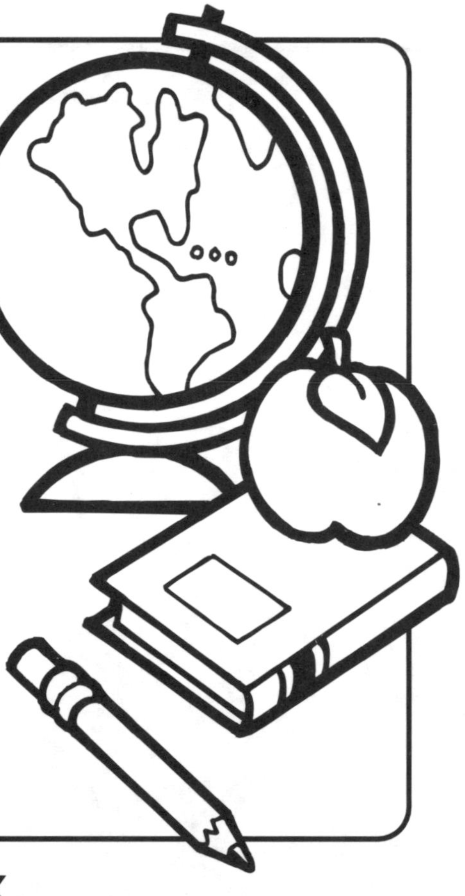

4

Time for work.

Hello Day! Hello Night! written by Rozanne Lanczak Williams

Hello evening! Hello night!

Hello Day! Hello Night! written by Rozanne Lanczak Williams

Time for play.

Time to go to sleep.

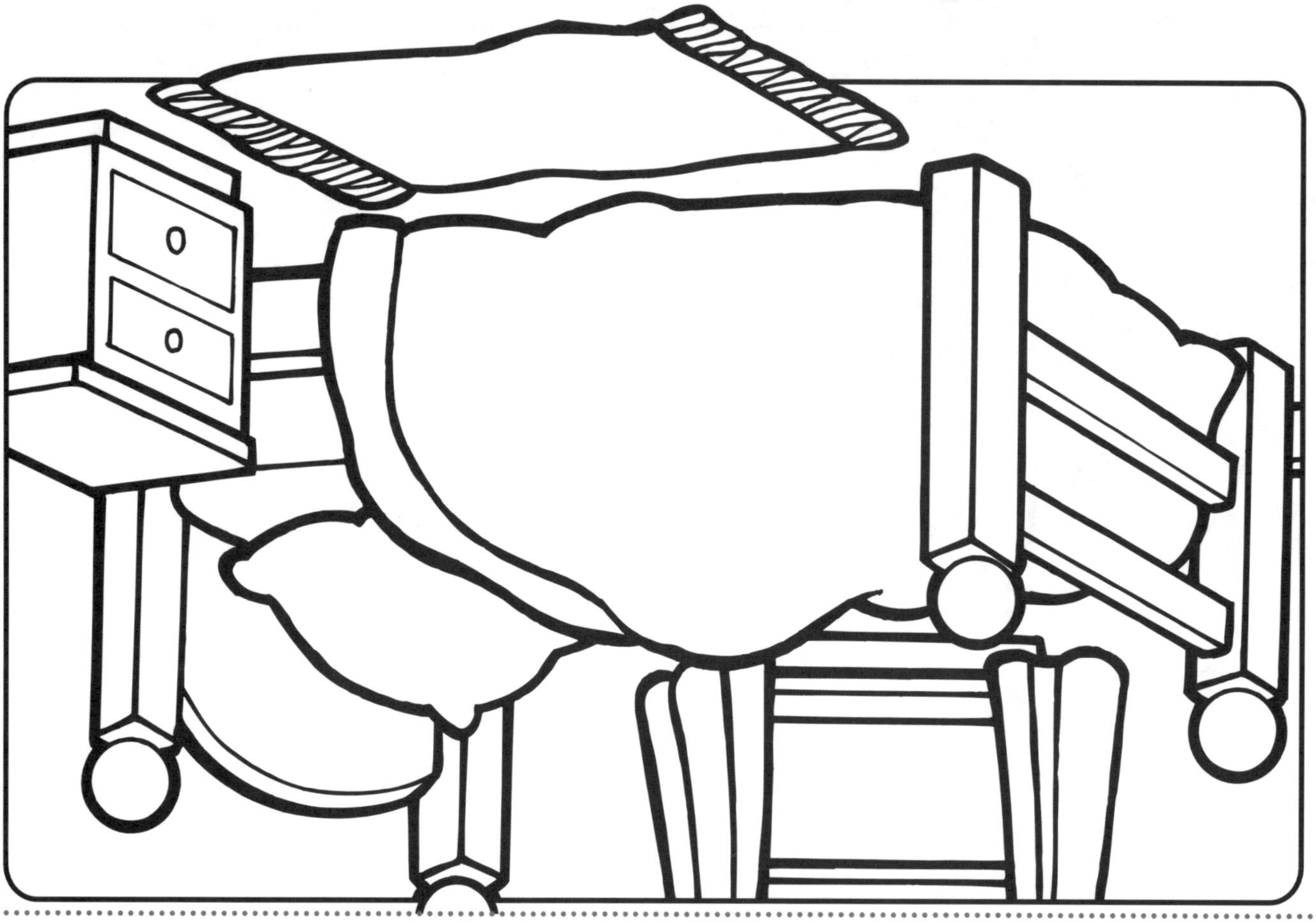

Good night!

Hello Day! Hello Night! written by Rozanne Lanczak Williams

A seed needs soil.

Potting Soil

What Does a Seed Need?

A seed needs water.

A seed needs air.

What Does a Seed Need? written by Rozanne Lanczak Williams

6

A seed needs time . . .

A seed needs sunlight.

5

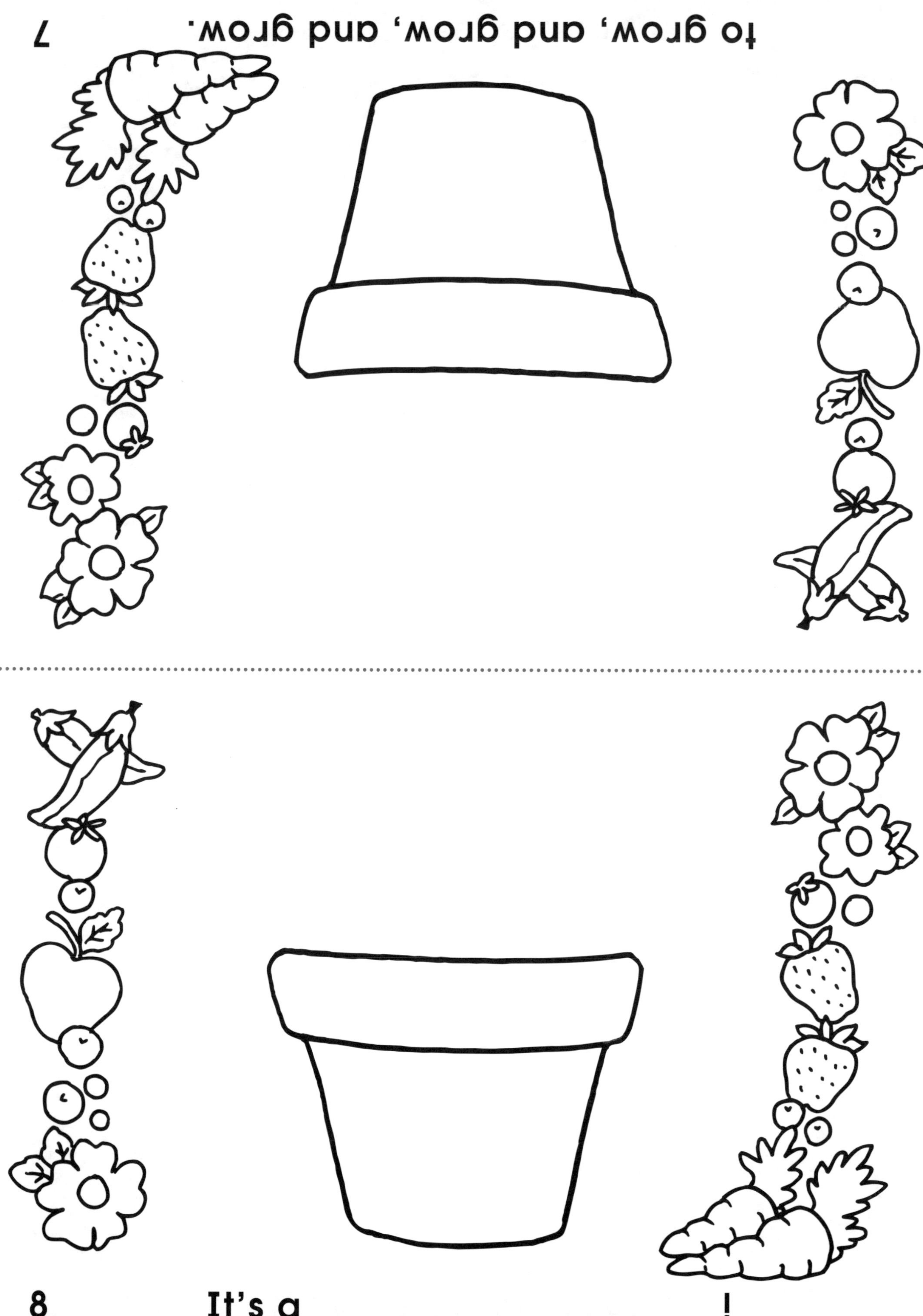

to grow, and grow, and grow.

7

8

It's a ______________________ !

What Does a Seed Need? written by Rozanne Lanczak Williams

Flowers grow.
2
Living Things Grow
1
Living Things Grow written by Rozanne Lanczak Williams
© Frank Schaffer Publications, Inc.

Trees grow.

3

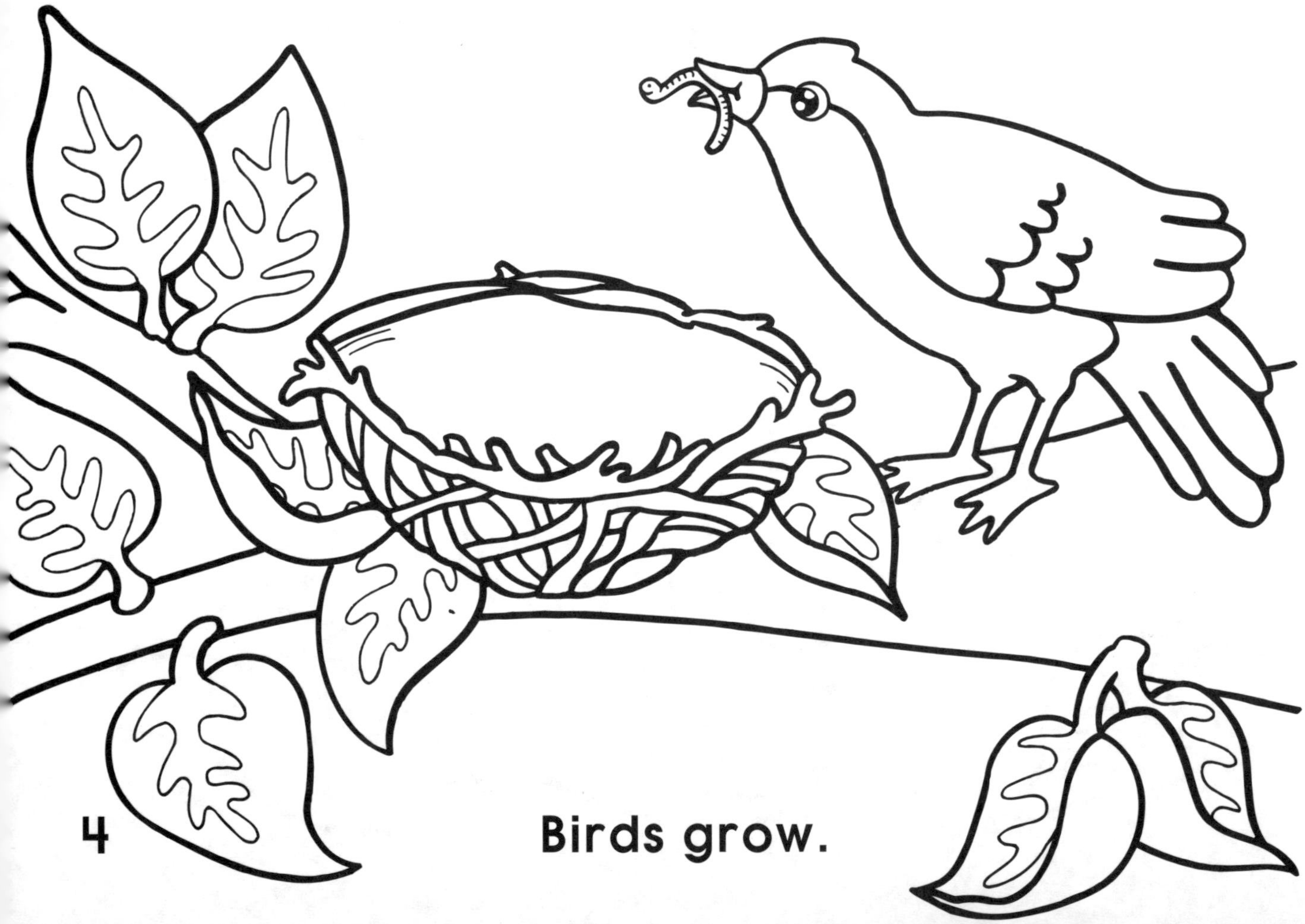

Birds grow.

4

Living Things Grow written by Rozanne Lanczak Williams

Living Things Grow written by Rozanne Lanczak Williams

Frogs grow.

What else grows?

A ________________ grows.

A ______________ grows.

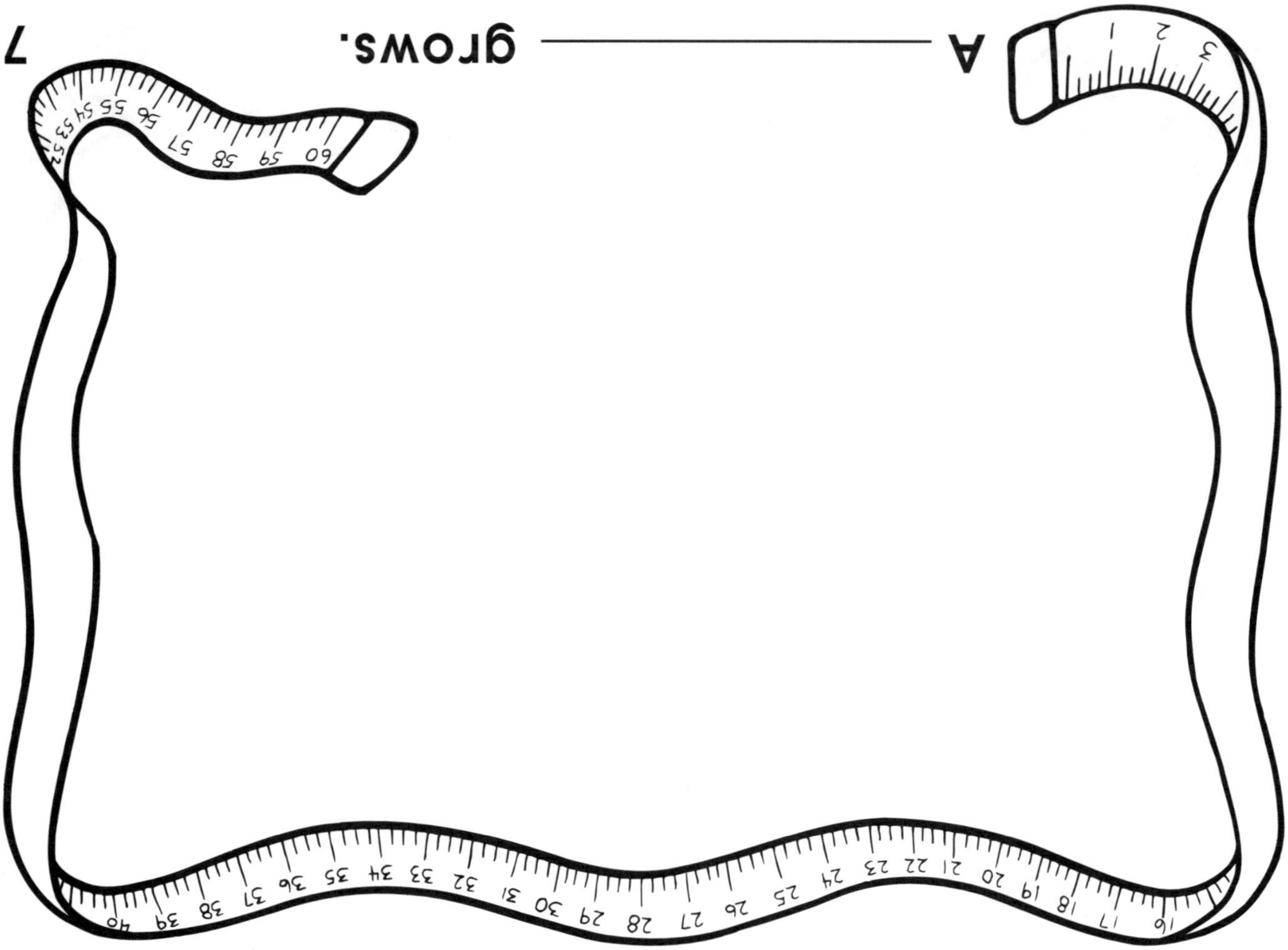

And I grow, too!

Animal Builders

A spider can build a web.

Animal Builders written by Rozanne Lanczak Williams

A bird can build a nest. 3

4 A beaver can build a lodge.

Animal Builders written by Rozanne Lanczak Williams

6 A ________ can build a ________.

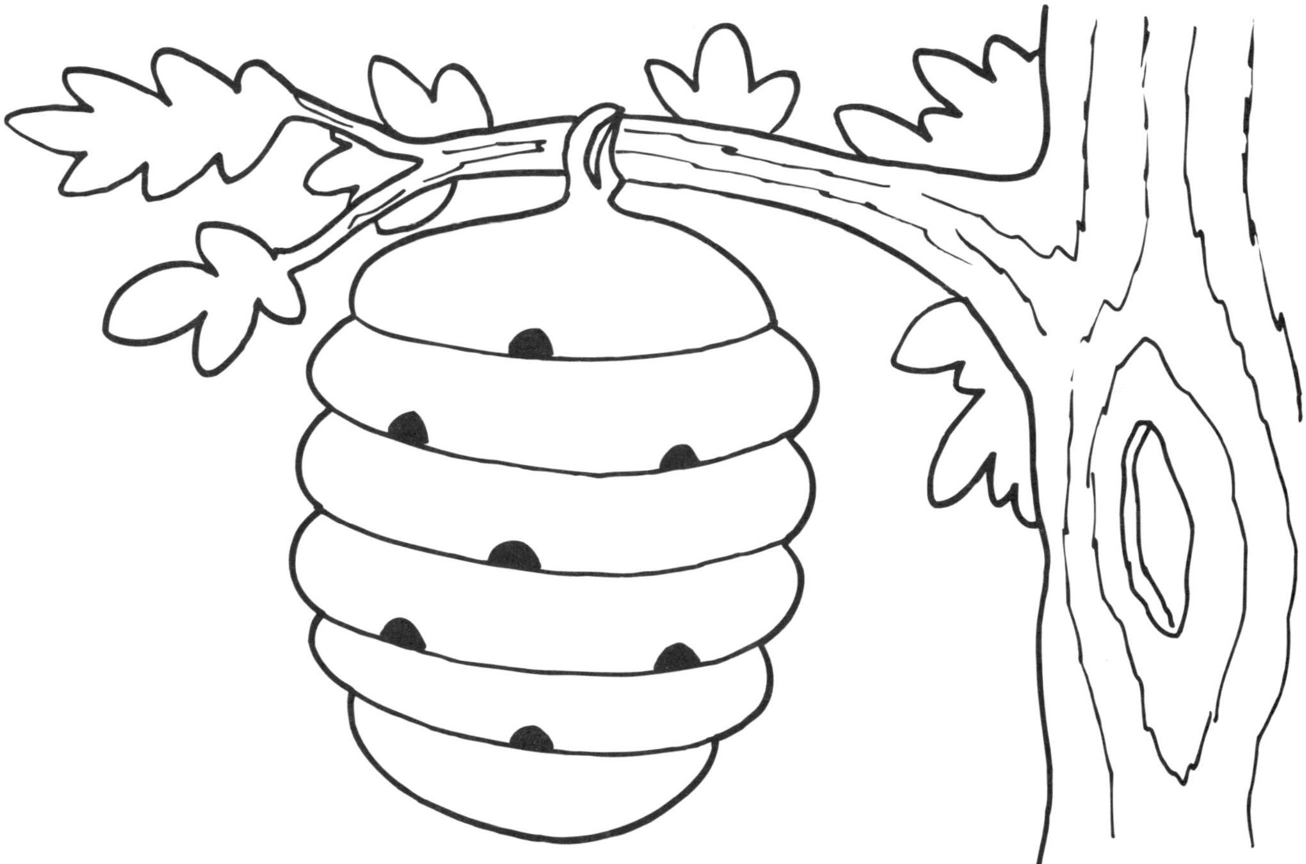

A bee can build a hive. 5

Animal Builders written by Rozanne Lanczak Williams

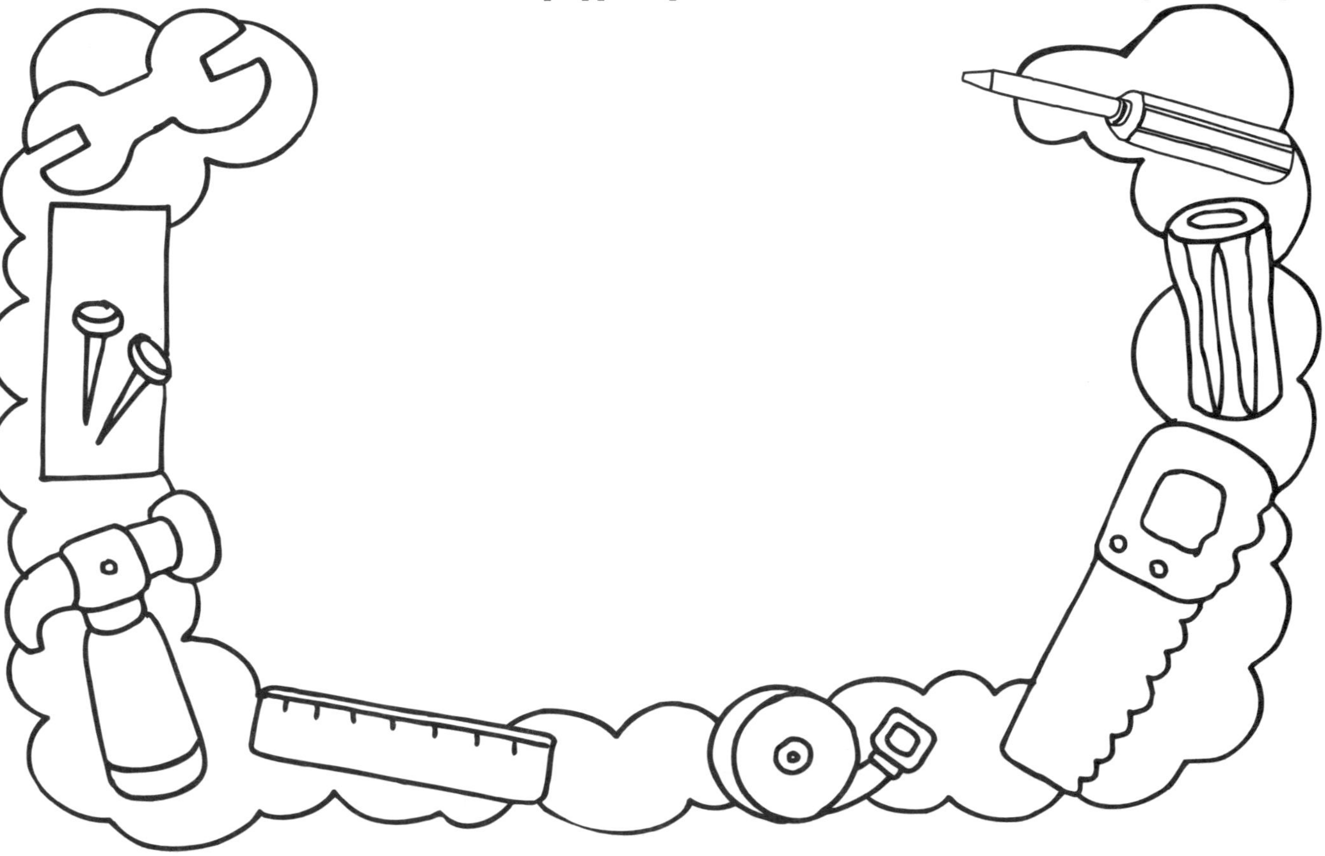

A __________ can build a __________.

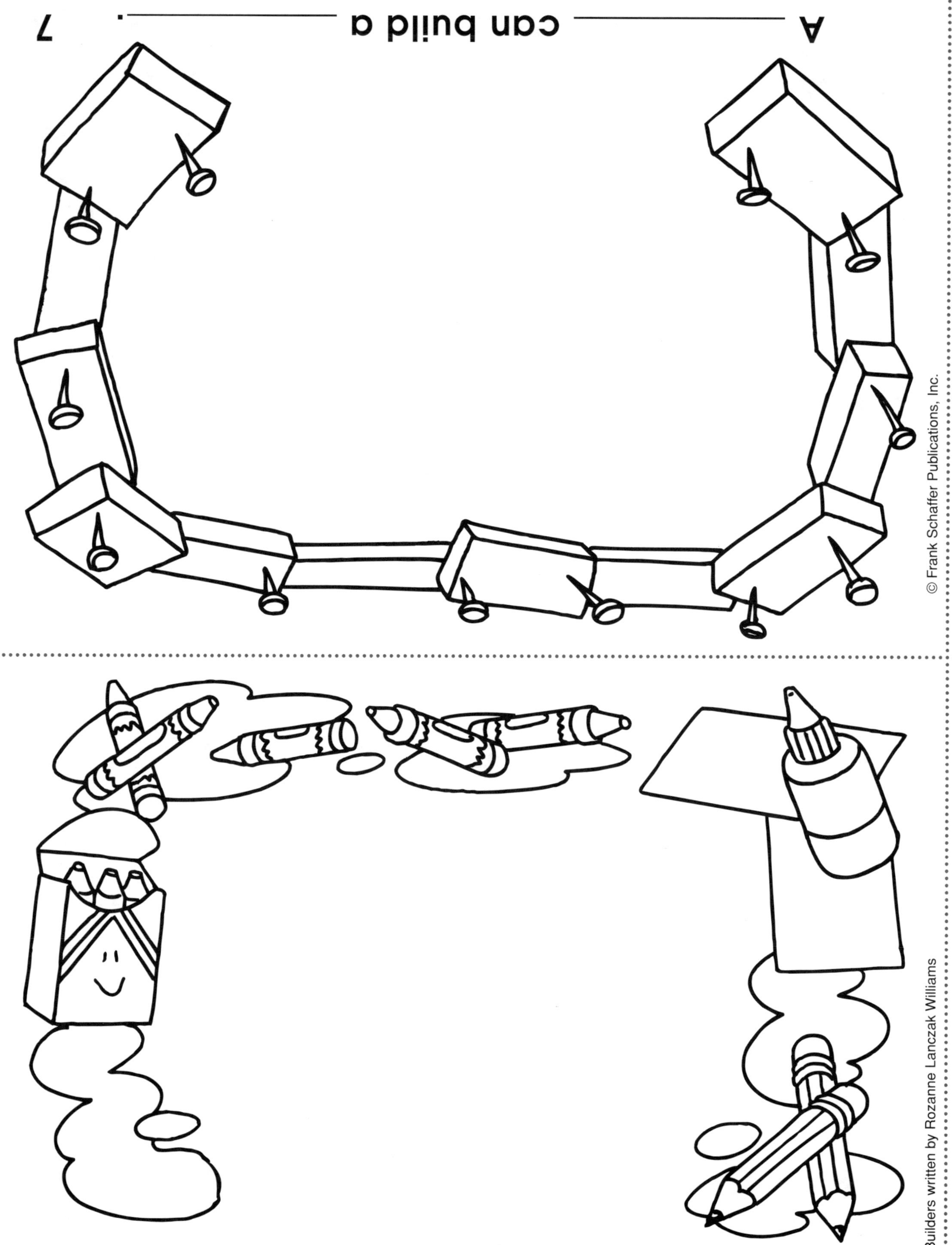

And I can build a __________!

Animal Builders written by Rozanne Lanczak Williams

2

I need energy to skate.

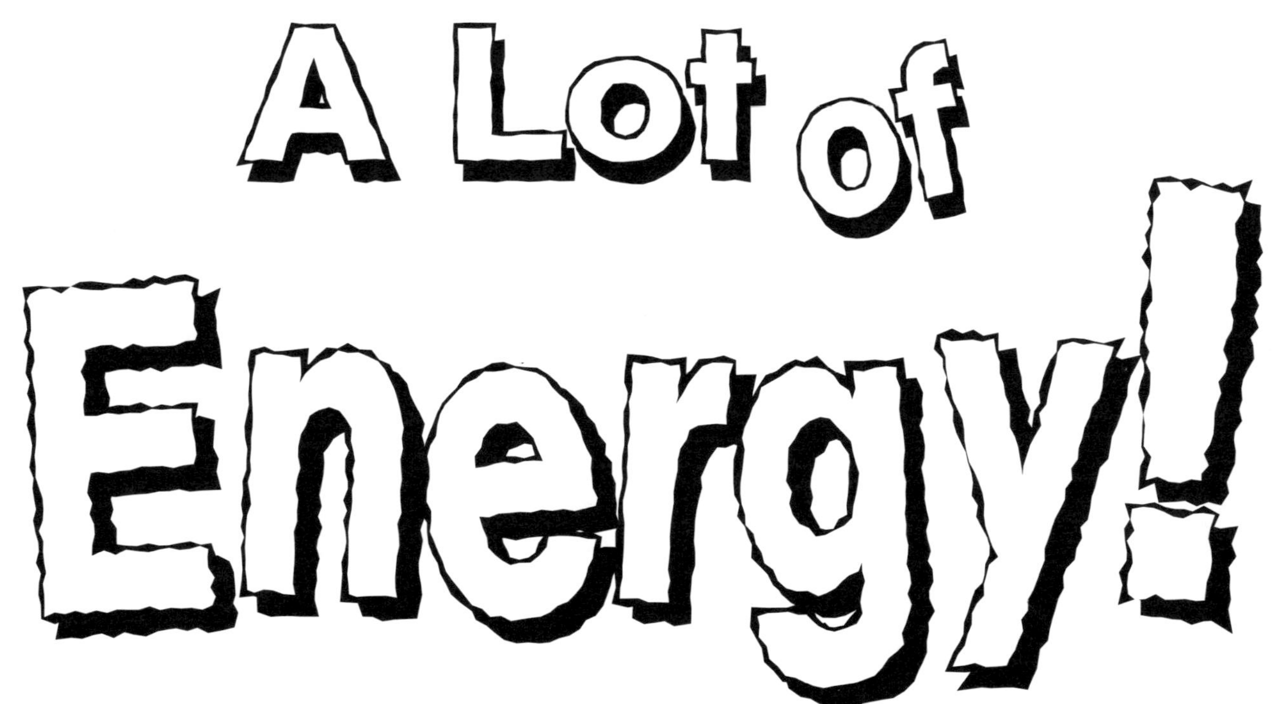

A Lot of Energy! written by Rozanne Lanczak Williams

3

I need energy to run.

4

I need a lot of energy

A Lot of Energy! written by Rozanne Lanczak Williams

I need energy to swim.

6

to jump rope in the sun!

5

I need energy to play.

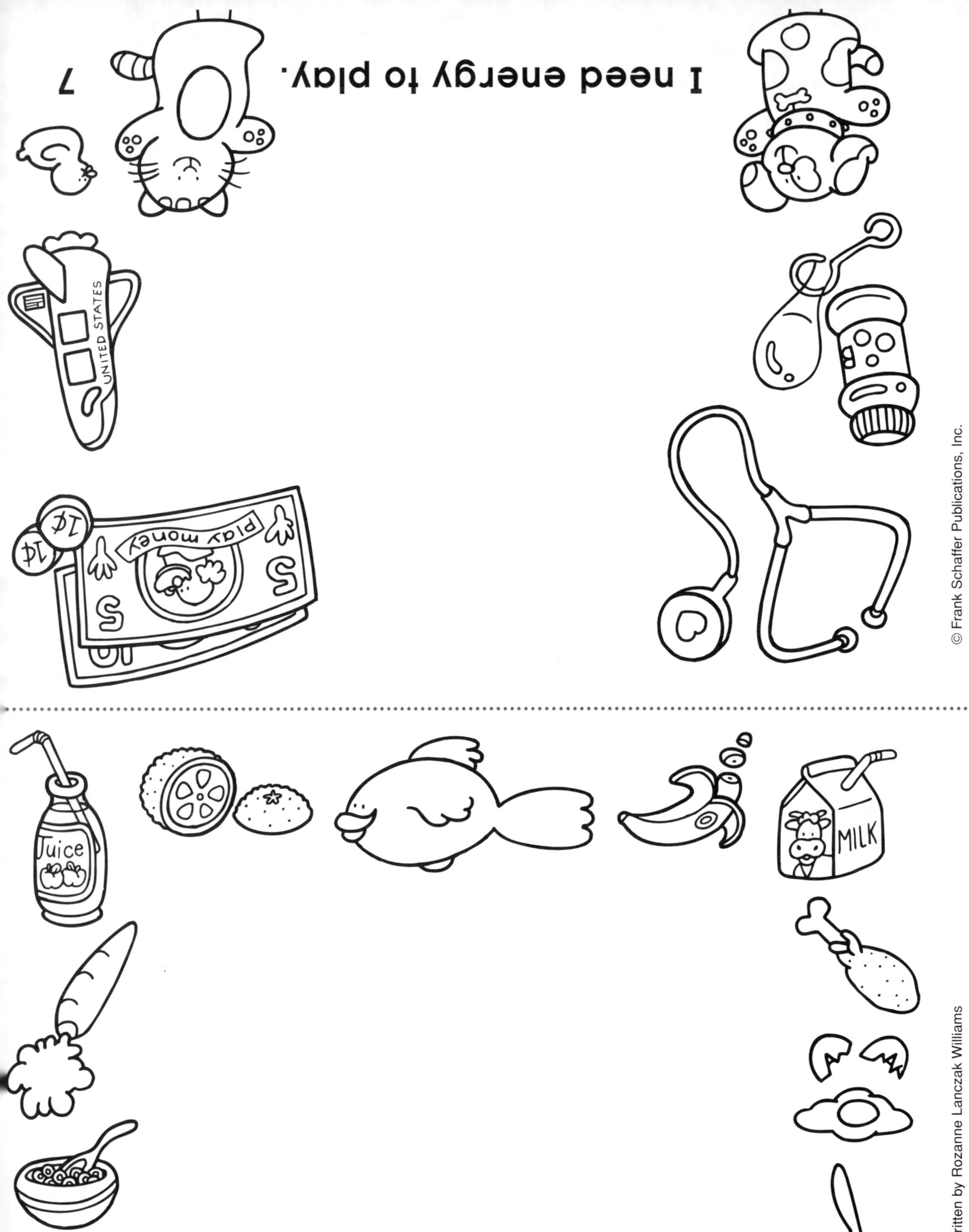

I get a lot of energy
from the foods I eat each day!

A Lot of Energy! written by Rozanne Lanczak Williams

2

I can see ______________.

I learn by seeing.

I Learn with My Five Senses

I Learn with My Five Senses written by Rozanne Lanczak Williams

I learn by listening.

I can hear .

3

I learn by smelling.

4 I can smell ________________.

I learn by touching.

I can touch ____________________. 6

I learn by tasting.

I can taste ____________________. 5

I learn about __________ by __________.

My five senses help me learn about my world.

What Am I?

I live in a pond. I have smooth green skin.
I eat insects. I am a ______________.

What Am I? written by Rozanne Lanczak Williams

I live in the desert.

I move along the ground.

I have no legs. I am a __________.

3

I live in the ocean.

I am the largest animal on Earth.

4 I have lots of blubber. I am a __________.

I live in the forest. I hunt at night.

I eat mice. I am an ________. 6

I live in the Arctic.

I am a bird, but I cannot fly.

I am a good swimmer. I am a ________. 5

What Am I? written by Rozanne Lanczak Williams

I live in a __________. I eat __________.

7 I can __________. I am a __________.

I live in a __________. I eat __________.

8 I can __________. I am a __________.

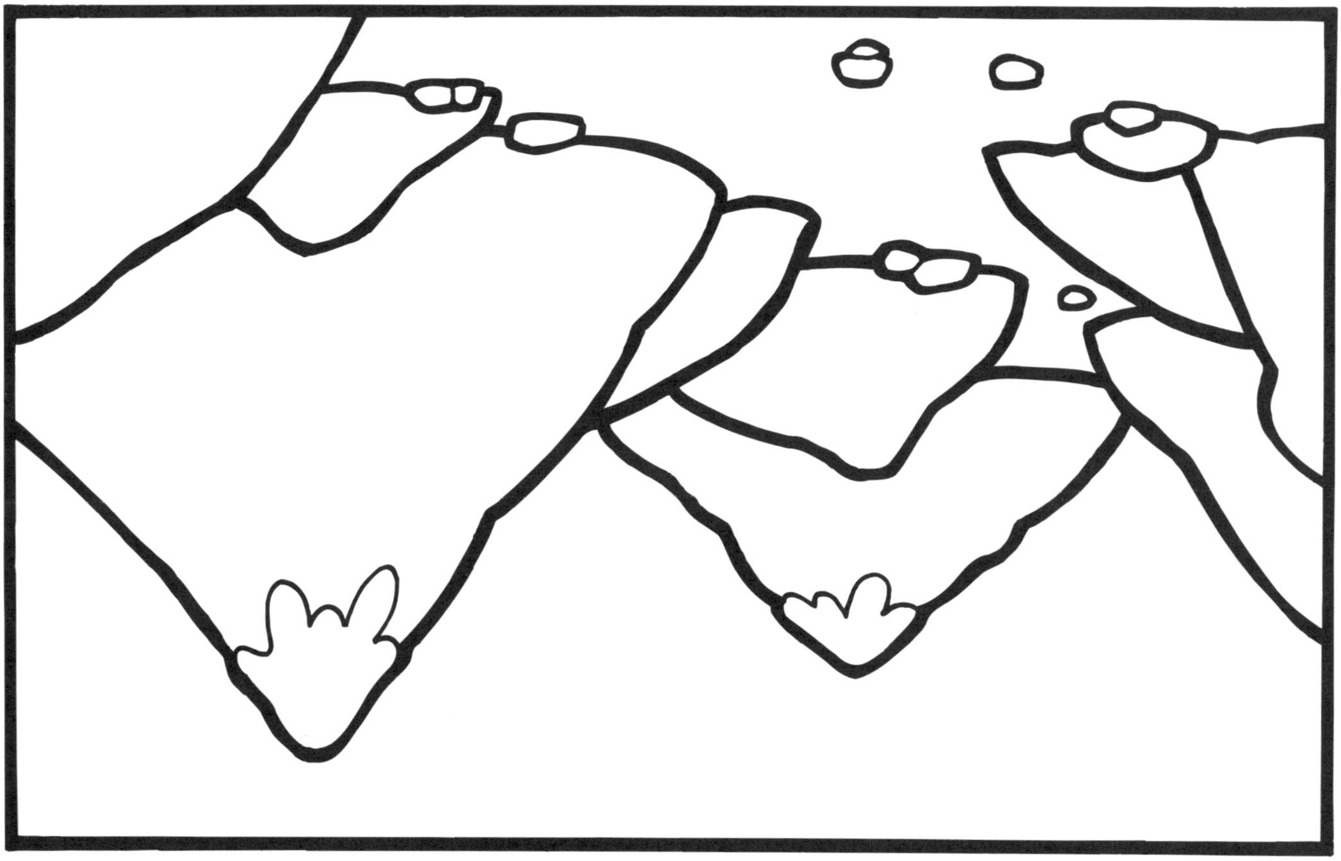

Hooray for the Earth!

Hooray for the mountains.

Hooray for the seas.

Hooray for the flowers.

Hooray for the Earth! written by Rozanne Lanczak Williams

Hooray for our planet.
Hooray for our home.

Hooray for the trees.

Hooray for the Earth! written by Rozanne Lanczak Williams

Let's take care of our planet.

Let's take care of our home!

Hooray for the Earth! written by Rozanne Lanczak Williams

I LIKE TREES

I Like Trees written by Rozanne Lanczak Williams

I like big trees,

little trees,

trees in the park.

I Like Trees written by Rozanne Lanczak Williams

I like trees in the daytime,

and trees in the dark.

I Like Trees written by Rozanne Lanczak Williams

I like spring trees,

summer trees,

trees in the fall.

I Like Trees written by Rozanne Lanczak Williams

Winter trees,
climbing trees—
I like them all!

Where Does Water Go? written by Rozanne Lanczak Williams

3 The rain fills the river.

4 The river flows to the ocean.

Water goes into the clouds.

Where Does Water Go? written by Rozanne Lanczak Williams

The sun heats the ocean.

7 The clouds get big and dark.

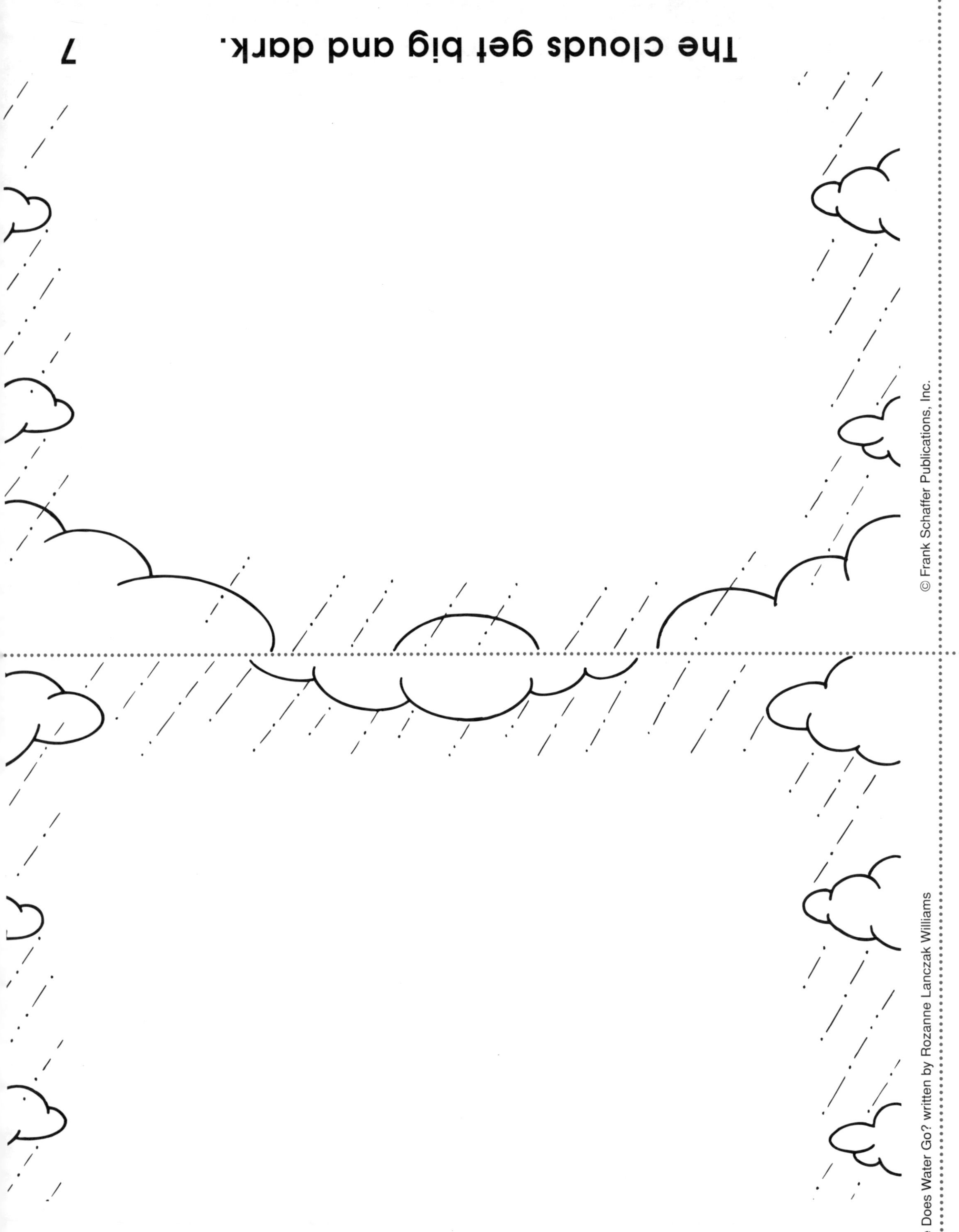

8 And down comes the rain again . . .

Where Does Water Go? written by Rozanne Lanczak Williams